My Underpants are Made from Plants

Where Stuff Comes From

By

Vera J. Hurst & Gloria G. Adams

Vera J Hurst
Creations 2019

Slanted Ink
www.slantedink.com

My Underpants are Made From Plants

Where Stuff Comes From

By

Vera J. Hurst & Gloria G. Adams

ISBN: 978-1-7367688-5-3

Vera J Hurst
Creations 2019

Slanted Ink

www.slantedink.com

Table of Contents

MY UNDERPANTS ARE MADE FROM PLANTS!

Underpants! Jeans! T-shirts! All of them are made from a plant called "cotton."

Cotton seeds live safe within a hard, round shell called a "cotton boll." Inside the boll, the seeds are cushioned by pillow soft fibers that pop out when the boll cracks open.

If you picked up a cotton boll and blew on it, the seeds would float away on the wind, carried by the cottony fibers.

Early farmers picked these fibers by hand and combed them to remove the seeds. The invention of a machine called a Cotton Gin made this faster and easier.

A single cotton fiber is about the size of a human hair and as long as a human finger.

Factory workers spin many of these fibers together into thread. They weave them into material and dye it different colors. Then they sew it into clothes, like T-shirts, jeans, and underpants!

Long ago, some people thought that cotton came from trees with lambs growing on the ends of the branches! They believed that when the lambs were hungry, the branches would bend to the ground to let them eat.

What else is made from
COTTON?

Q-tips	Dollar bills
Bandages	Tents
Towels	Sails for boats

THERE'S BEE SPIT IN MY HONEY

That sweet, thick honey on your bread really is made using bee spit!

Honey is not just a yummy treat for us to eat. Bees eat it too! Honey gives the bees energy and food to survive cold winters.

Honeybees gather syrup, called nectar, from flowers. They store the nectar in a special stomach called a "honey sac." When a bee flies back to the beehive, he moves this nectar from his honey sac to his mouth. Then he uses his tongue to "spit" it to another bee.

That bee places the nectar from his mouth into a small hole called a "cell." Many of these cells make up a Honeybee's hive. Other bees fan the cell with their wings. This turns the nectar into thick honey.

Finally, more bees cover the opening of the cell with a thin layer of wax called "beeswax."

Much of the honey today comes from flowers called clover. It takes the nectar from a million of these flowers to make just one pound of honey!

What else is made from
HONEY AND BEESWAX?

Lipstick	**Medicine**
Lotion	**Soap**
Candles	**Shoe Polish**

MY PAJAMAS COME FROM WORMS

Would you wear clothes made from worms? That is where the fabric called silk comes from!

Silkworms start their lives as butter-colored eggs as small as the head of a pin. When they grow bigger, they wrap themselves inside a cocoon. The cocoon is made from a sticky liquid that squirts out of the mouth of the silkworm and turns into a solid thread.

One thread can be as long as 1500 baseball bats laid down end to end!

Silk farmers soak each cocoon in water and carefully unwrap the thread. It is twisted together with other threads at a factory and made into a soft fabric.

Silk looks shiny and smooth like a slippery eel, but it is not slippery at all. In fact, silk is so strong it can be used to make bullet-proof vests!

Silkworms help the environment. They eat the leaves from Mulberry trees, which grow where other plants can't grow. This helps farmers use more of their land.

What else is made from
SILK?

Parachutes	Curtains
Sheets	Ribbons
Stockings	Stationery

THERE'S OIL IN MY CRAYON

Can black liquid oil be made into a solid, colorful crayon? Yes!

The rainbow colors in your box of crayons began as a thick oil called "petroleum."

Petroleum oil is pumped up from deep below the ground. Special chemicals separate a sticky wax from the oil. This wax is called "paraffin wax."

To make crayons, paraffin wax is melted and heated until it's very hot. Colored powders are added. A machine shoots the hot wax into crayon-shaped molds.

Then the molds are plunged into cold water. This makes the crayons harden in just a few minutes. Machines wrap labels around each crayon. The crayons are packed into boxes and brought to stores.

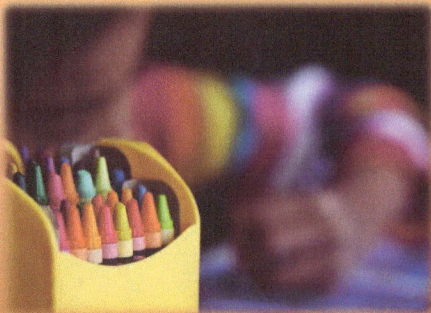

The Crayola factory makes 8,500 crayons every minute. That's over 4,000 boxes of crayons every hour!

The first box of crayons had only eight colors. Today, Crayola makes 120 different colors! What a colorful world it is now, and it all began with gooey black oil.

What else is made from
PARAFFIN WAX?

Candles   **Chewing gum**
Wax Paper   **Waterproofing**

THERE'S SAND IN MY EYEGLASSES

When you run your hand over the surface of a piece of glass, it's very smooth. But that piece of glass actually started out as a gritty pile of sand.

To get glass from sand, machines mix the sand with limestone and soda ash. Brick ovens heat it up so hot that it melts and looks like thick honey. When it cools down, the glass hardens, and other machines form it into long sheets.

Different shapes are cut from the sheets. Big pieces can make windows and doors and small pieces can be used to make eyeglasses. Glass can also be poured into molds or shaped into bottles or jars.

The first eyeglasses were called "reading stones." They were big pieces of glass that made things look bigger when you looked through them.

Even though some eyeglasses are still made from glass, today most of them are made from plastic.

Glass is not new. Thousands of years ago, the Egyptians made glass beads and used them for jewelry.

What else is made from
GLASS?

Aquariums   Mirrors

Marbles   Windows

Light Bulbs   Solar Panels

THERE ARE VEGGIES IN MY T-SHIRT

Soybean plants can be as small as a tulip plant or as tall as a basketball player. They have long fuzzy stems that produce hairy bean pods. There are 60-80 pods on a plant and each pod contains three soybeans. That is like having a bag of jellybeans on every plant!

At the factory, soybeans are ground up, cooked, and made into a food called "tofu." The thick water left over from making tofu is pushed through machines. What comes out looks like long strings of spaghetti.

These strings are dried and woven together to make a very soft cloth. The soybean cloth is sewn into different kinds of clothing, like T-shirts, hoodies, and underwear.

Soybean Pods

Henry Ford, who was a car maker, was the first person to wear a suit and tie made from soy fibers. He even used material made from soy fibers to line the door panels in his cars!

Soybeans are sometimes used to make crayons. Just one acre of soybeans can produce 82,368 crayons.

What else is made from
SOYBEANS?

Soy sauce   Milk
Yogurt   Ink
Candles   Fuel

MY SODA CAN IS MADE FROM ROCKS

Did you know that crumbly rocks and dirt can be turned into smooth, round soda pop cans?

Pop cans begin as a rock called bauxite that is made up of a mixture of minerals. One of these minerals is called alumina, which can be turned into aluminum.

It takes about four tons of bauxite to produce one ton of aluminum. That's as heavy as 16 hairy gorillas!

Bauxite is dug from the ground and placed into giant crushing machines. This creates white powder that looks a lot like salt. Then it takes a hot bath in some chemicals and gets zapped with electricity. The material that sinks to the bottom is pure aluminum.

The hot aluminum is cooled and formed into big sheets, rods, or blocks. The sheets are cut into pop can shapes and baked in huge ovens. One factory can produce 1500-1800 cans every minute!

Aluminum can be reused! After it's melted, your pop can could end up as part of an airplane wing!

13

What else is made from
ALUMINUM?

Kitchen pans	**Aluminum foil**
Cookie sheets	**Ladders**
Picture Frames	**Baseball bats**

MY COAT COMES FROM A GOAT

**Coats from goats?
Socks from an ox?**

Many of the clothes you wear started out as the furry coats of animals!

To make clothes from animal coats, a farmer must give the animal a haircut. This is called "shearing." The fluffy coat that has been shaved off is called a "fleece." The fleece has to be washed and combed. Then, machines spin it into long strands of yarn. People use yarn to knit clothes like hats, socks, and mittens.

Weaving the strands together makes a solid fabric. Sheep give us a very warm fabric called "wool." We also get fabric from the fleece of alpacas, camels, goats and llamas.

One of the softest yarns in the world comes from the undercoat of a musk ox. It is called "qiviut." Fuzzy "angora" yarn is made from the fur of Angora rabbits.

Sheep even help make baseballs. Baseball makers wind 219 yards of wool yarn inside each baseball. If you laid the yarn out in a straight line, it would be longer than 2 football fields!

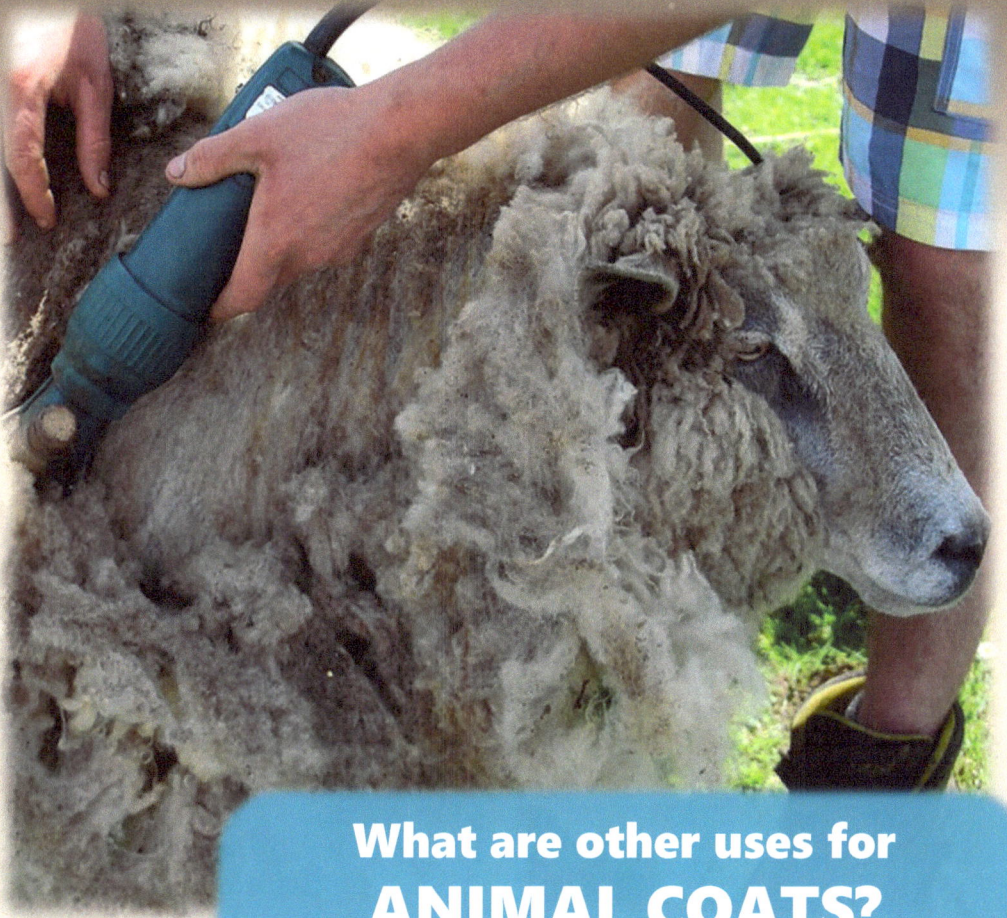

What are other uses for
ANIMAL COATS?
Sweaters
Carpeting
Furniture cloth
Blankets

MY CHOCOLATE BAR IS FULL OF BEANS

How can smooth, thick chocolate come from a bunch of beans?

The small dark brown beans of the cacao tree are called cocoa beans. They are what make your chocolate candy bar so yummy. The beans grow inside seedpods shaped like small footballs. There can be 20, 30, or even 40 cocoa beans inside each pod.

The seedpods are chopped from the trees. Then the beans are dug out of the gooey pulp. Factories roast and crush the beans into a thick syrup. The syrup is heated up, poured into molds, and cooled down. Sugar, milk, and vanilla are added.

Factories use molds to turn chocolate into different shapes and sizes. A special mold makes the famous Hershey® Kisses. The factory turns out 70 million Hershey® Kisses every day!

Some chocolate companies have created giant chocolate bars. The biggest one so far weighs almost 13,000 pounds!

Long ago, chocolate had other uses, like medicine and even money. At one time, people could buy a rabbit for just ten cocoa beans.

What are other uses for
COCOA BEANS?

Beauty creams   **Lip balm**
Fertilizer   **Mulch**

THERE ARE BURPS INSIDE MY BREAD

Did you know that bread is full of gas?

A loaf of bread begins as a seed that grows into a plant. The seeds from this plant are called grains.

There are many different kinds of grains, such as wheat, rye, oats, barley, rice, and millet. When they are ripe, the seeds are picked and ground into flour.

Bread dough is made by mixing the flour with water and a tiny organism called yeast. Yeast is what makes bread burp!

Yeast eats the natural sugar in grains and produces gas-filled bubbles. These bubbles grow bigger, making the bread dough "rise". When bakers make bread, they have to let it rest for a while. This gives the yeast time to eat the sugar and make the bubble burps.

Yeast is not a new discovery. Scientists say that yeast was used to make bread in Egypt 6,000 years ago!

The next time you eat a sandwich, take a close look at your bread. The tiny holes are where the gas bubbles "burped," making the bread soft and spongy.

19

What else is made from GRAINS?

Cereal	Donuts
Noodles	Pancakes
Tortillas	Pizza

THERE IS ELEPHANT POOP IN MY PAPER

Would you write on paper made from elephant poop?

People all around the world use this type of paper every day. But don't worry; the poop is washed first!

When animals eat plants, the fibers are broken down when they go through an animal's stomach. The tough fibers left in the poop are used for making paper.

Workers wash the poop to kill germs. They boil it with baking soda and salt to get rid of the smell. Then they mash it into tiny pieces called pulp.

They mix the pulp with other plant fibers, like cotton and linen. Then they strain it and flatten it into long sheets. It is dried and sometimes color is added. The result is "recycled" writing paper!

People have found other uses for animal poop. One company makes it into a special kind of coffee. There is even jewelry made from dinosaur poop!

An elephant can poop about 10 gallons of poop a day.
This makes about 115 pages of "poop" paper.

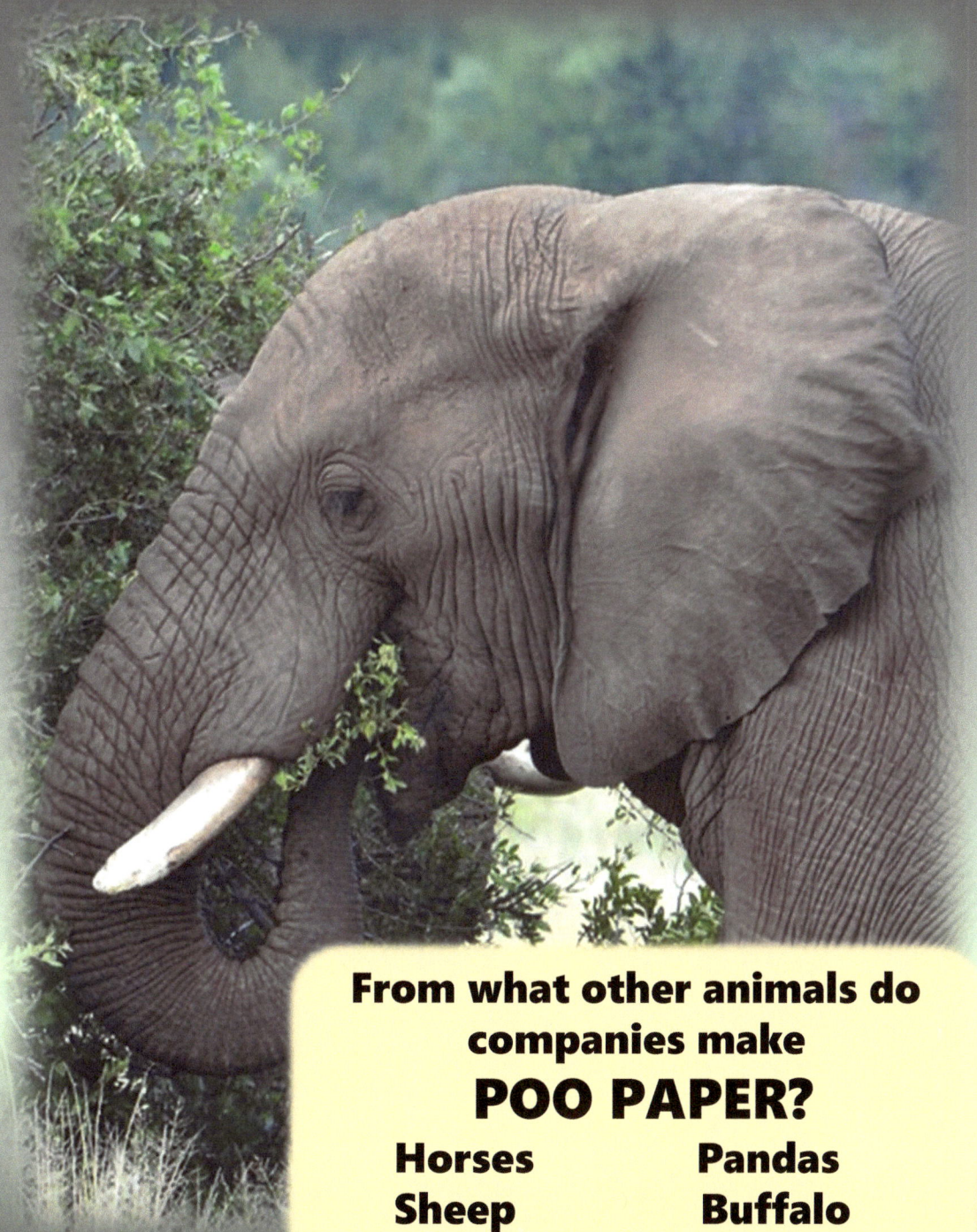

From what other animals do companies make

POO PAPER?

Horses   **Pandas**

Sheep   **Buffalo**

Giraffes   **Moose**

THERE'S TREE SAP IN MY BIKE TIRES

Did you know that your sturdy, tough bike tires started out as white goo that drips from a tree?

That goo is the sap inside rubber trees. Rubber trees are very tall and grow in warm rainforests. The soft, creamy sap that forms just under the tree bark is called "latex." It is the same color as milk.

Workers make cuts in the rubber trees. This is called "tapping." The sap runs out of these cuts into small cups. Some workers can tap the latex from 400 trees in just one day!

The latex is shipped to factories. Workers mix it with special chemicals. This turns it from a white goo to hard rubber fiber. The fiber is shaped into blocks. Big rolling machines press the blocks into thin sheets.

These sheets are cut into different shapes, like tires for cars, airplanes, trucks, and your bike!

Something else is made from rubber: rubber ducks! The biggest rubber duck ever made is 40 feet tall. That's as tall as four houses piled on top of one another!

What else is made from
RUBBER?

Raincoats	Rubber bands
Boots	Erasers
Balloons	Sneakers

Explore Some More!

Check out these websites on how things are made:

How Products are Made: http://www.madehow.com

How Stuff Works: http://science.howstuffworks.com

Videos about how things are made:
https://www.weareteachers.com/how-things-are
-made-videos/

Glossary

Cocoon: a covering wrapped around a silkworm to protect it

Baking soda: a powdery white chemical

Fiber: a thin thread from a plant

Grain: the seeds from plants used for food

Limestone: a hard rock formed from shells or coral

Mineral: a solid, non-living substance found under the ground

Mold: container into which liquid is poured to create a given shape when it hardens

Nectar: the sweet syrup inside flowers

Paraffin wax: a soft wax made from oil

Petroleum: a kind of oil that comes from below the ground

Pulp: the inner juicy part of a fruit or vegetable

Sap: a watery juice inside a tree

Seedpod: the part of a plant that has seeds inside

Shearing: shaving off the fur or hair- from an animal

Strand: something that is long like a string

Tofu: a solid food made from soybeans

Yeast: a small living thing that makes bread dough rise

SELECTED
BIBLIOGRAPHY

Websites

"Baseball." How Products Are Made. Np., n.d. Web 6 Nov. 2019. <http://www.madehow.com/Volume-1/Baseball.html.>

"HowStuffWorks "Science"." HowStuffWorks. N.p., n.d. Web. 18 Nov. 2013. <http://science.howstuffworks.com/>.

"Soy clothing - softest fiber of all." Soy clothing - softest fiber of all. N.p., n.d. Web. 19 Nov. 2013. <http://www.naturalclothingcompany.com/soy>.

"The Story of Chocolate." The Story of Chocolate. N.p., n.d. Web. 18 Nov. 2013. <http://thestoryofchocolate.com/>.

"The Aluminum Association." The Aluminum Association. N.p., n.d. Web. 19 Nov. 2013. <http://www.aluminum.org/>.

Books

Murray, Julie. *Sheep to sweater*. Edina, MN: ABDO Pub. Co, 2007. Print.

Nelson, Robin. *From flower to honey*. Minneapolis: Lerner Publications Co., 2003. Print.

Oxlade, Chris. *Cotton*. Chicago, Ill.: Heinemann Library, 2002. Print.

Polin, C. J. *The story of chocolate*. New York, NY: DK Pub., 2005. Print.

Silk. White River Junction, Vt.: Nomad Press, 2011. Print.

Slavin, Bill, and Jim Slavin. *Transformed: how everyday things are made*. Toronto: Kids Can Press, 2005. Print.

Sohn, Emily, and Lisa Klobuchar. *Fabric: it's got you covered!* Chicago, Ill.: Norwood House Press, 2011. Print.

Wool. White River Junction, Vt.: Nomad Press, 2011. Print

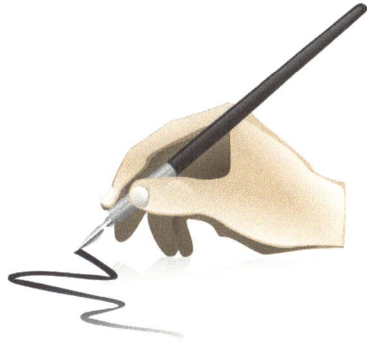

ABOUT THE AUTHORS

Vera J. Hurst is a poet, photographer and birder who enjoys traveling and hiking. Her works include *21st Century Trains,* a non-fiction, STEM-based book for fourth graders, published in 2018 by Enslow Publishing, and *Seven Deadly Sins: Simply Delectable Stories,* in 2016, published through Third Thursday Publishing, LLC.

Gloria G. Adams spent most of her career as a children's librarian. She is the co-author of *Ah-Choo!,* a children's picture book, written with Lana Wayne Koehler. She has been published by Sterling Children's Books, Enslow, Rosen, Greenhaven Press, and The Children's Writer's and Illustrator's Market. Check out her website: www.gloriagadams.com.